NUMBER 468

THE ENGLISH EXPERIENCE

ITS RECORD IN EARLY PRINTED BOOKS
PUBLISHED IN FACSIMILE

The publishers acknowledge their gratitude to
the Trustees of the Britisch Museum for
their permission to reproduce the
Library's copy, Shelfmark: 529.g.6 (3)

Library of Congress Catalog Card Number:
75-38202

S.T.C.No. 13701
Collation: A-E4

Published in 1972 by

Theatrvm Orbis Terrarvm Ltd.,
O.Z.Voorburgwal 85, Amsterdam

&

Da Capo Press Inc.
-a subsidiary of Plenum Publishing Corporation-
277 West 17th Street, New York N.Y. 10011

Printed in the Netherlands

ISBN 90 221 0468 0

The vse of the two Mathematicall instruments, the crosse Staffe, (differing from that in common vse with *the mariners :*) *And the Iacobs Staffe:* *Set* foorth Dialogue wise in two briefe and plaine Treatises: the one most commodious for the Mariner, and all such as are to deale in Astronomicall matters: the other, profitable for the Surueyor, to take the length, height, depth, or breadth, of any thing measurable.

Set forth by Th. Hood. mathematicall Lecturer in the Citie of London.

Newly reviewed, and the second time imprinted.

The Staues are to be sold in Marke lane, at the house of FRANCIS COOKE.

Imprinted at London by Richard Field for Robert Dexter. 1596.

TO THE RIGHT HONO-RABLE, CHARLES LORD HOWARD, BARON OF EFFING-*HAM, LORD HIGH ADMI-rall of England, &c.*

WHen I presented this little booke in writing vnto you (right Honourable) I did little thinke that it should haue come vnto the Printers hand: but the Staffe being shewed vnto diuerse men that were desirous to know the vse thereof, I thought the labour would be too long to teach euerie one, and to write it out continually, the pains would haue bin much more. Therefore to saue both my paines in writing, and my labour in teaching, I could finde no better meanes, then to commit it vnto the presse. Considering then that it must abrode, I beseech your Honour, that it may appeare vnder your name, that as it pleased you at the first to giue it a friendly entertainment, so now you would also be Patron vnto the same, that your authority protecting, and your wisdome allowing it, you may stay the vnaduised censure of some rash conceipted men. Wherby I shalbe both encoraged to a further study, and bound to pray for your Honours increase in all good things.

Your Honors most humble,

Thomas Hood.

To the courteous Reader.

I Present vnto thee (gentle Reader) in one ſmall Volume, the vſe of two Mathematicall Inſtruments, the one is the Croſſe Staffe: the other, the Iacobs Staffe. In the deliuery of their vſe, I haue done mine endeuour to lay out the matter moſt plainely, that thou maiſt finde commodity in the thing. That is all my deſire, which if I may attaine vnto, I ſhall be glad. If any thing ſeeme otherwiſe vnto thee, accept my good will, whoſe purpoſe was to ſatisfie thy minde: and interprete hereof fauourably, at the leaſt, becauſe it is my firſt labour that euer came abroad. So ſhall I be the more willing to take farther paines, and to commit ſuch other things vnto thy ſight, as I haue in hand. Farewell.

Thine *Th. Hood.*

A Dialogue touching the vse *of the Crosse Staffe.*

Scholler.

HAuing vnderstoode that you haue a kinde of Crosse Staffe, somewhat differing from those of the common sort, I am somewhat bold to request a litle conference with you about the same.

Maister. Not so bold as welcome: It is my desire to further all men what I can. I pray you therefore briefly to propound what you would haue, and I will returne an aunswer to your contentation, so farre forth as I may.

Sc. First, I desire to know, what end you haue generally in the thing.

Ma. The Staffe may be vsed to diuerse purposes: but my chiefest end and intention therein, is to take the height of the Sunne aboue the Horizon.

Sc. May not a man do it as well by the other Staffe, commonly which is called Bella Stella?

M*a.* Comparisons are odious, therefore I omit that matter, and request you to propound some other question. Yet this I say, that this Staffe is not so troublesome as that, which you haue named. Moreouer, the degrees thereof are greater then those of the other Staffe, though the Staues themselues be of equall length. For 5. of them are as bigge as 15. in the Bella Stella. Neither are you tied vnto an Horizon: for it may be vsed both by land and sea: and without applying it vnto your sight.

Sc. These be no small commodities. If therefore I may

obtaine at your handes the vse of this Staffe, I shall thinke my selfe not a litle beholding vnto you.

Ma. Your request is graunted: Take therefore what order you thinke best for your instruction.

Sc. Because I would haue all things plainly deliuered, I will keepe this order in my questions. First, I wil begin with the two peeces of wood: and then with the sockets: thirdly, I will request the vse of the Vane: and last of all, howe the Sunnes altitude may be found out.

Ma. Do so: and I will labour to fulfill your mind.

Sc. First therefore, what names doe you giue vnto these two peeces of wood?

Ma. I call the one the Yard, the other the Transame.

Sc. What meane these markes that are in them?

Ma. They are degrees of altitude, with their minutes.

Sc. But how shall I distinguish the Transame from the Yard, seeing the degrees in them both be all of one quantitie, neither hath the one of them any more degrees then the other?

Ma. It is true that there be 45. degrees in each of them. The Transame containeth the first, namely, all the degrees from 1. to 45: the Yard containeth the rest, namely, the degrees from 45. vnto 90.

Sc. Is that of necessitie that they should do so?

Ma. No: But by this meane I make the degrees so great as they are.

Sc. Then I perceiue this: that from the rysing of the Sunne, vntill he be raysed 45. degrees aboue the Horizon, his eleuation must be taken in the Transame. Againe, after that he is 45. degrees aboue the Horizon, his eleuation must be taken in the Yard.

Ma. It is so.

Sc. May I easily knowe by this staffe, when the Sunne is vnder 45. degrees, and when he is aboue 45. degrees of eleuation?

Ma. Most

Ma. Most easily, and readily: as you shall be taught hereafter.

Sc. Why is the yard cut off so close vnto the 90. degrée?

Ma. That it may the better come vnto your sight: wherein the Centre of the circle is imagined from whence all the lines do come.

Sc. Thus much concerning the Transame, and the yard. Now for the Sockets: to what purpose serue they?

Ma. They be made to ioyne the Transame and the yard together.

Sc. What principall note is to be obserued in them?

Ma. They must be ioyned together squire wise, at right angles.

Sc. To what end are the notches in them?

Ma. The notches serue to this purpose, to see how to set the Transame, and the yard iust vppon his place. For the brasse carying a certaine thickenesse with it, doth hinder my sight so, that I cannot well iudge of the true place whereon they should stand, were it not for these notches, which must bee no déeper, then the thickenes of the brasse, or if they be déeper, they must be both of one depth.

Sc. To what end serue the Scrues?

Ma. They serue to none other purpose but this, namely, to kéepe the Sockets fast in their places that they slip not.

Sc. May I put the Sockets vpon the yard which way so euer I will, without any difference?

Ma. No: But you must take héede of this, that the Socket wherein the Transame must be put, must alwaies hang downeward, for so it is best: otherwise there is not any great héede to be taken. And yet for auoiding of confusion, let this be your rule, that The end whereat your two Sockets close together, be next vnto your sight.

Sc. To what end serueth the Vane?

Ma. It serueth to take the eleuation of the Sunne, and also to saue your sight from the beames thereof.

Sc. In what maner I pray you?

Ma. Of that I will enforme you hereafter: for the maner being diuers, cannot wel be ſet down here, without confuſion.

Sc. Whereupon muſt the Vane ſtand?

Ma. Sometimes vpon the Tranſame, and ſometimes vpon the Yard, according as the manifold vſe of the Staffe requireth.

Sc. Is it any matter to which hand it ſtandeth? either to the right hand, or to the left?

Ma. Yea that it is: and this Rule ſhal alwayes ſerue your turne, When the Tranſame ſtandeth on the right hand of the Yard, let the Vane bend toward the left hand: and contrariwiſe, when the Tranſame ſtandeth on the lefte hand of the Yard, let the vane bend toward the right hand.

Sc. Now let me know how I ſhall ioyne the Tranſame and the Yard together: for to that end were the brazen ſockets made.

Ma: They are to be ioyned together diuerſly, according as the Staffe is diuerſly vſed.

Sc. Is the ſtaffe then to be vſed after more waies than one?

Ma: Yea, for you may chooſe whether you will apply it vnto your eys (as you do the Balla ſtella) or whether you will holde it in your hand, and find the height of the Sunne by the ſhadow of the Vane.

Sc. Is there any generall obſeruation to be noted in theſe ſeuerall vſes?

Ma. Yea: For this is a generall rule, that whether you apply it vnto your eye, or whether you holde it in your hand, the Tranſame muſt be alwaies perpendicular vnto the Horizon, and the yarde leuell vnto the ſame.

Sc. But how ſhall I be aſſured of the perpendicularitie and the parallelitie of the one and the other?

Ma. At the Sea, where your Horizon is frée, if you bring the Yard euen with the cloſing of the heauen & the Horizon, it is enough: But vpon the Lande, or looking out at a Window of an house, or standing in the streete, you may adioyne vnto the one side of your Transame a Plummet of Leade, thereby to finde out his perpendicularitie. Yet do I not thinke that so necessary, because I know, that a little practise will make the eie a reasonable iudge both of the perpendicularity, and of the parallelitie.

Sc. Let vs now come vnto the particular vse of the Staffe. You say that in seeking out the eleuation of the Sunne, I may choose whether I will set the Staffe vnto mine eie (as I do the Bella stella) or whether I will hold it in mine hand.

Ma. It is so.

Sc. I pray you therefore instruct me first concerning the vse of that, when I must set it vnto mine eie.

Ma. This one thing is not to be neglected, that it must stand on the ball of your cheeke, for that place is most conuenient, because it doth not lightly yeelde, neither is the sight thereby distorted: Yet if any man thinke it meeter to apply the Staffe to the corner of his eie next to the eare, I will not gainesay him, because the sight is not to euery man so ready in one place as it is in an other.

Sc. How must I in this case ioyne, the Transame and the Yard together.

Ma. Put the Yard into his socket, so that the notch may stand iust vpon the 45. degree. Then put the Transame into his socket in like manner, that the notch may be vpon the 45. degree.

Sc. Is it not materiall which waies the degrees stand, either vpward or downeward, or on the side?

Ma. Let this be your rule: Keepe the degrees of the Yard vpward, and let the degrees of the Transame be toward you, so shall you be the lesse deceiued. Otherwise to them that are cunning, it is neither here nor there.

Sc. Is there any thing to be obserued in the placing of the Vane, more then hath beene said before?

Ma. This is to be noted, that the edge of the Vane muſt be placed iuſt with the higheſt degree of the Tranſame.

Sc. Each thing being fitted in this maner, how ſhall I take the height of the Sunne?

Ma. Firſt, you muſt know whether the Sunne bee more or leſſe then 45. degrees aboue the Horizon.

Sc. How ſhall I know that quickly? for me thinketh that ſhould be troubleſome.

Ma. It is moſt eaſie. The Tranſam and the yard being ioyned together, as you haue béene taught: turne the 90. degrees of the yard toward your breaſt (it is no matter for ſetting it vnto your eie) and bring the Tranſame right againſt the Sunne, vntill the ſhadowe thereof fall euen with the ſide of the yard. If the ſhadow of the Tranſame from the toppe of the Vane downward be longer then the yard, it is a moſt certaine ſigne that the Sunne is vnder 45. degrees high If it be ſhorter, the Sunne is more then 45.degrees high.

Sc. If the Sunne be leſſe then 45. degrées high, what muſt I do?

Ma. Set the Staffe cloſe to your eye, kéeping the yard parallele vnto the Horizon, as you haue beene taught before: then let the Tranſame ſinke downe into his ſocket, vntill the toppe of the Vane come euen with the Centre of the Sunne, the number of the degrées from the top of the vane to the notch of the Socket will declare his eleuation.

Sc. This thing is plaine enough. May not his eleuation yet be found out any other way by the Tranſame, while the Sun is vnder 45. degrées.

Ma. It may be found out otherwiſe in this manner: Put that end of the Tranſame into his Socket, which conteyneth the beginning of the degrées, let the firſt degrée come euen with the notch, as the 45. degrée did before, and make the Tranſame faſt with the ſcrue: then ſet the yard vnto your eie, and kéeping it leuel, moue the Vane vp and downe, vntill the vppermoſt edge thereof come euen with the Centre of the

Sunne

Sunne: the degrée vppon which it lighteth is the Sunnes altitude.

Sc. Thus much concerning the Sunne whiles he is vnder 45. degrées. What must I do to finde his altitude being aboue 45. degrées?

Ma. The Yard, Transame, and Vane being orderly placed, as it was taught before, set the yard vnto your eie, and kéeping it leuel, draw the Transame to you ward, vntill the Vane be euen with the Centre of the Sunne: the degrée then of the yard, vpon which the notch of the socket lighteth, declareth vnto you the height of the Sunne.

Sc. Is there any more to be sayde concerning the vse of the Staffe, when it is applied vnto the eie?

Ma. No: For I haue at large declared the same: so that now it remaineth to tell you, how the Staffe may be vsed, although it be not applied vnto the eie.

Sc. One thing let me first demaund: May not I take the height of any starre in the same manner as I haue taken the height of the Sunne?

Ma. Yes, that you may: Neither is there any difference in the one from the other. You may also take the distaunces of any two starres thereby, so that they be not further asunder then 90. degrées, following the selfe same precepts which you haue bene taught concerning the Sunnes altitude For if the two stars be lesse then 45. degrées asunder, they will be within the end of the Transame, and the yard. Then set the yard against the one starre, and draw the Transame into his socket, vntill the edge of the Vane light vpon the other star, and the degrées of the Transame will shew their distaunce. If the two starres propounded be more, then 45. degrées asunder, it is certaine, that they will be without the compasse of the Transame. Then draw the Transame towarde your sight, vntill the end of the yard light vpon the one starre, and the edge of the Vane vppon the other starre, and the degrées of the yard will shew their distaunce, so that you may take the distance of any two starres that are 90. degrées asunder.

Sc. Now let vs returne vnto the former question: how must the Transame and the Yard be ioyned together, and how must the Vane be placed, if I be not disposed to set the staffe vnto mine eie?

Ma. The ioyning of them, is after diuerse maners, according as the vse doth differ. First, you may find the sunnes eleuation in the same maner, as you did before, although you do not apply the staffe vnto your eie, and therefore in this case the Transame and the Yard must be ioyned together, as they were before, and the Vane must also stand after the same manner.

Sc. How then shall I finde the Sunnes eleuation?

Ma. Set the palme of your hand close to the end of the Yard where the 90. degrée is placed, and then turne the Transame vpon the Sunne, and pull downe the Transame into the socket, vntill the shadow of the toppe of the Vane fall iust vppon the end of the Yard close to your hand: Then the degrée of the Transame next vnto the socket, is the Sunnes eleuation.

Sc. I perceiue that in this case also I must hold the yard very leuell.

Ma. Yea, that you must: and so shall you be sure of the Sunnes altitude.

Sc. May I finde the Sunnes eleuation thus, if he be more then 45. degrées high?

Ma. Yea, euen as you did before. For this is all the difference, that your hand standeth now in stead of your eie And I am of this minde, that if you take héede vnto the leuell of your yard, you may take the Sunne as precisely this way, as you do with applying the staffe vnto your eie. Besides, this is one commodity, you shall not dazell your sight, and hurt it as you did before.

Sc. What other way is there yet to vse the Staffe, and how must I ioyne the Transame and the yard together?

Ma.

Ma. There is yet an other way to finde out the height of the Sunne without setting the staffe vnto your eie. For the performaunce of which conclusion, you must set the yard and the Transame together thus. Put the Transame into his socket so, that the beginning of the first degrée may fall iust with the notch: then make it fast with the scrue: put the yard also into his socket, in such sort that the 45. degrée fall euen with his notch, and set the Vane at the other end of the yard, as you were instructed before.

Sc. How must I vse it then?

Ma. You must here also haue that consideration which you had before of the Suns eleuation, whether he be more or lesse then 45. degrées, which is knowen by the rule giuen in the beginning.

Sc. If the Sunne be lesse then 45. degrées high, how must I hold the staffe?

Ma. Let the Transame hang downeward toward your féete, and turne the Vane which is vpon the yard to the sunne vntill the shadow of ẏ yard fall iust with the side of the Transame: the degrée of the Transame, vppon which the shadow o the Vane doth fall, declareth the height of the Sunne.

Sc. Is it materiall which side of the shadow I do chose?

Ma. Yea, For that side of the shadow must be chosen, which is made by that part of the Vane, which is next vnto the end of the yard, as you may sée in the figure.

Sc. I perceiue your meaning in this. But how must I hold the staffe, if the Sunne be more then 45. degrées high?

Ma. Then must the Transame lie leuell with the Horizon, which before did hang perpendicular, and the yard must stand vpright toward your Zenith. Then turne the yardes toward the Sunne so, that the shadow thereof fall close to the side of the Transame, marke vpon what degrée of the Transame the shadow of the vppermost side of the Vane falleth, subduct that out of the 90. degrées, and the remainder sheweth the Sunnes eleuation.

Sc. May not the height of the ſtarres bee found out in this manner?

Ma. No: By reaſon their beame is ſo weake that it cannot caſt a ſhadow; but the beame of the ſunne being ſo ſtrong, his height may thus be eaſily found out, and very readily, and alſo certainely, hauing onely this care, that the ſtaffe be kept leuell as it ought to be.

Sc. Is there any thing elſe to be ſaide concerning the vſe of this ſtaffe?

Ma. It may be vſed to diuerſe other purpoſes (as I ſaid in the beginning:) namely, to the deſcription of Regions, and Countries: to the finding out of the quantities of the Eclipſes of the Sunne and Moone: with the longitude and latitude of the Planets. This ſtaffe alſo ſerueth to the correcting of the places of the fixed ſtars: & to the ſetting downe of the amplitude of the riſing or ſetting of the Sunne or Stars. But mine intent was onely to take the Sunnes eleuation therwith. Yet haue I told you how to take the altitude of the ſtars with their diſtances, by the ſame ſtaffe: which for this time ſhall ſuffice. The other propoſitions I will lay forth, either in my publike lectures which I read, or at other more conuenient time, as my leyſure serueth.

FINIS.

THE VSE OF THE IACOBS STAFFE.

Imprinted at London by Richard Field, for Robert Dexter. 1596.

TO THE RIGHT HONORABLE, L. IOHN LVMLEY, BARON OF LVMLEY. &c.

AFter that I had penned (Right honorable) the vſe of the Croſſe ſtaffe, I was deſired by mine acquaintance, to take the paines in the Iacobs ſtaffe. I did ſo: my paines were liked and called (as the former) vnto the preſſe. Without a Patrone I would not haue it come forth, becauſe I would not be thought to bee without a Patrone. To rrouble one man with them both, I thought it not good, leaſt I ſhould ſeeme forgetfull of other my wellwillers: among whom your good will, ſhewed vnto me at diuerſe times, hath emboldened me to craue your patronage for the thing, which if it ſhall pleaſe your Honour to vndertake, I ſhall accompt it as a ſpeciall token of your louing fauour, and acknowledge my ſelfe in recompence, bound to pray for your Honours daily encreaſe in all kindes of bleſſings.

Your Honours moſt
humble. Th. Hood.

A Dialogue touching the vſe of *the Iacobs Staffe.*

Scholler.

SInce my laſt departure from you, I haue diligently applied the croſſe Staffe, according vnto the rules which you haue preſcribed, and finde the ſame to be very conuenient for the purpoſe, which you haue ſet downe: that is, for the taking of the height of the Sunne and Starres, with their diſtances: Yet can I not ſtay my ſelfe here, but muſt once againe preſume vpon your courteſie.

Maiſter. What is it that you deſire? I loue not many words but you ſhall find me moſt ready to pleaſure you, or any man els, as far forth as my ſimple skill will afford me.

Sc. This is the matter: I would gladly learne the vſe of the Iacobs Staffe as well as I haue learned the vſe of the croſſe ſtaffe: May I not therefore ſo deuide the croſſe ſtaffe on the one ſide, that it may ſerue my turne for a Iacobs ſtaffe, ſo that I be not troubled with the cariage of any more then one?

Ma. Yes, very well. And loe, here is one ſo diuided, that as the one ſide of the Yard and Tranſame ſerueth for the Sun and Starres, ſo the other ſerueth for any dimenſion.

Sc. This falleth out euen as I would haue it: ſeing therefore the Staffe is ready at hand, if your leyſure ſerue, I pray you inſtruct me in the vſe thereof.

Ma. With a good will: Moue you the queſtions, and I will anſwere them.

Sc. Why doe they call it Iacobs Staffe? Was he the firſt inuentor of the thing.

Ma. I know not that: but they take occasion to call it so, by reason of those wordes which are written Gen. 32. 10. where the Patriarch sayth, *That with his Staffe he came ouer Iordane:* Wherein I thinke, they misconstrue his meaning. Notwithstanding, by whom soeuer it was inuented, the Instrument questionlesse is of singular vse.

Sc. Let that passe, and declare I pray you, what affinity this Staffe hath with the other. For comparisons will make a thing more plaine.

Ma. First, as in the other Staffe, the degrées of the Yard and Transame are all one in number, so are the partitions in this. Secondly, as the yard and Transame in that, did ioyne together squarewise and at right angles, so must they doe in this. Thirdly, there must the same consideration be had in placing of the Vane in this, that was had in the placing of the Vane in the other: that is, it must stand euen with the beginning of the partitions, except some speciall occasion require otherwise, as is to be séene in the measuring of breadthes. Fourthly, One of the two, either the Yard or the Transame, as occasion serueth, must be parallele vnto the thing measured, the other must be perpendicular.

Sc. Is that a rule to be noted in the Iacobs staffe?

Ma. Yea that it is: otherwise it is most certaine you commit an errour.

Sc. Then haue I béene amisse informed, as I haue séene men, that haue thought them selues to be of no small cunning in the vse of the Staffe, who measuring the height of a thing, haue made no matter how they held their Staffe, so that they did sée the toppe and the bottome of the thing to be measured, euen with the two ends of their Transame.

Ma. Beléeue me in this, that it cannot be that their practise should be true. This one argument shall serue for a thousand. You know that the proofe of the vse of the Iacobs staffe standeth vppon the proportion of Triangles. Now those Triangles onely are proportionable which are like, and in like maner

ner situated, which they can neuer be except either the yard, or the Transame be parallel or perpendicular (as occasion serueth) to the thing measured.

Sc. Your reason is good that it shoulde be so: but me thinketh that the perfourmaunce thereof should bee impossible.

Ma. Why so?

Sc. You know that many Fields be full of Dales and hollow places: others be full of Hilles and Mountaines. Moreouer, many things stand not vpright vpon the ground, but leane either forward or backeward, or else to the one side or to the other. Last of all in the bredth of a thing there is a great variety. For be it in a wall, or fielde, or what else whatsoeuer, there is for the most part either hollownesse or roundnesse, or some such like thing which may make it impossible in my iudgement to draw either parallele or perpendicular line vnto them.

Ma. I graunt those things to bee as you say: yet doe they not hinder a man from taking the measure of their length, height, or breadth. For where the place it selfe doeth not affoorde a leuelnesse: it is an easie matter to imagine it leuell. Thus also you must thinke with your selfe, that although Geometry bee the Arte of measuring well, and generally nothing may be excluded from the compasse thereof, yet must we not bee too curious in particulars, least we shew our selues either ouer wise, or else too foolish. The Phisitian generally professeth to heale all diseases, but when he commeth to deale with particular bodies, he faileth in his art, and yet is not to be discommended so long as he doeth that which Art requireth. It is enough for the Geometrician, if he giue the iust length of a streight line drawen betweene two places assigned. Or if hee can not doe it by one streight line, he may doe it by many. Whosoeuer immediately (for mediately he may finde out the length of a crooked line by meanes of a straight line) whosoeuer, I say, requireth any more of him then that, he abuseth both the man and the art. Therefore, when any thing is propounded to be measured, be it length, height, or breadth, you are first of all to consider, whether the thing lie leuell, or stand vpright, or be straight, as it is required that it should be. If it be straight, you may presently fall in hand to measure it, as your rule shall direct you: If it be not straight, but be either crooked, or leaning one way or another, it must be your principall care, first to fit the same to your purpose: as for example, in measuring the distance of a thing (for I will first begin with that dimension, and will afterwards teach you what you shall doe in the rest) if the distaunce betwéene two places assigned be not straight, as you sée in these figures which follow it is not, then choose out two points in the distance, the one being at the one end thereof, and the other in the other, betwéene which you may imagine a right line to be drawen, and then so hold your Staffe, that the yard or the Transame may be parallele, or perpendicular vnto it, as occasion shall serue.

Sc. You haue throughly satisfied me in this. Tell me now wherein these staues do differ, séeing you haue declared wherin they doe agrée.

Ma. Their difference consisteth in two things especially: in their partitions, and in their vse.

Sc. Touching the first difference, it is euident indéed, that the parts of the other Staffe are vnequall, and that this is diuided into equall portions: but maketh it any matter howe many the parts are into which this Staffe is deuided?

Ma. No: The more the better, and more precise will your worke be. And therfore, if you could diuide them into 100000. it were the more commendable, but 1000. sufficeth in this Staffe, which though it be not deuided into more, yet you may imagine it to be an hundred times as much as it is.

Sc. What is the difference in the vse of the Staues?

Ma. The generall difference is this: the other serueth for Astronomicall matters: this serueth for measuring of land.

Sc. What

Sc. What meane you by thoſe words, Meaſuring of land? Is it to take the breadth, or length only of a field? or may it be applied otherwiſe?

Ma. The truth is this, that by this Staffe we may meaſure the length, the height, and the breadth of any thing, all the which we expreſſe by thoſe words, Meaſuring of land: hauing relation to the word Geometry, which ſignifieth the ſame if it be conſtrued word for word.

Sc. Before I doe come vnto the particular meaſuring of theſe thrée dimenſions, length, height, and breadth, what generall notes haue I to obſerue?

Ma. Firſt, that whether the Yard or the Tranſame be applied vnto your ſight, it muſt ſtand cloſe vnto the ball of your cheeke.

Sc. But which end of the Yard or Tranſame muſt ſtand to the ſight ward?

Ma. That which hath the beginning of the degrées: for ſo it is moſt conuenient for the readineſſe of your account. The ſecond note is this: to winke with one eie. For the ſight by that meanes will aime at your marke the better, when it is not parted but brought into one. The third obſeruation is this: that, the diſtance of your marke be not ouer great.

Sc. Why do you giue this note, conſidering that in the vſe of the other Staffe, there was no limitation made of any diſtance? For you meaſured therewith the ſtarres in heauen.

Ma. The caſe is not all one, becauſe the Starres haue a brightneſſe which helpeth their ſight, and their beame commeth directly from the Centre of their body, vnto our eye. But in terreſtriall things it is not ſo: For in them, by reaſon of the weakeneſſe of our ſight, we may be greatly deceiued, except they be within a reaſonable compaſſe.

Sc. What do you call a reaſonable compaſſe, or diſtance?

Ma. It cannot well be defined, becauſe there are ſo many things required therein, that they will hardly iumpe all together generally: as theſe, the cleareneſſe of the aire: the goodneſſe of the ſight: the largeneſſe of the inſtrument. The aire may be cleare, but the ſight of a man may be dull. Againe, the ſight may be quicke enough, but the aire may be thicke. Againe, both the aire and ſight may ſerue the turne, but the inſtrument which he is to vſe may be ſo ſmall, that he can not take any thing farre diſtant from him. So that we cannot generally determine of the diſtance otherwiſe then thus, ſaying that that is a reaſonable diſtaunce, which is proportionable both to our inſtrument and ſight.

Sc. Procéede I pray you in the matter wherein you were.

Ma. The fourth regard muſt be had vnto your hands that they be ſteddy. For otherwiſe you may miſſe of the proportion of your meaſure. The laſt note is this, to take heede vnto the place wherein you make your ſtation.

Sc. How ſhall I be ſure of that?

M*a.* Thus: You know this, that either you ſtand vpright, or elſe leane one way or other. To be ſhort, howſoeuer your ſituation be, whether it be ſtanding, ſitting, leaning, or vpright, that is the place of your ſtation, which is right beneath your eie, ſo directly as is poſſible to be iudged and taken.

Sc. Theſe fiue things you ſay are generally to be noted. 1. The ſetting of the Saffe at the ball of the eye: 2. The wincking with one eie: 3. A reaſonable diſtance: 4. A ſteddy hand: 5. The certaine place of your ſtanding. Let vs now come vnto the particular vſe of the Staffe in the thrée dimenſions. You ſay, the Jacobs Staffe ſerueth to take the length, the height, and the breadth of any thing: Are theſe dimenſions found out any more waies then one?

M*a.* Yea: The length, and the height, are found out thrée ſeuerall waies, and the breadth two waies. The firſt and the ſecond way require one ſtation: the laſt way in each dimenſion is performed by two ſtations.

Sc. What meane you by a ſtation? and what is it to performe a thing by one or moe ſtations?

M*a.* A ſtation is nothing elſe but the place wherein you

ſtand:

stand: to performe a thing therefore at one station, is nothing else but by a standing in one onely place by the helpe of your instrument, to take the measure of a thing. To performe it at two stations, is by standing in two seuerall places to take your measure.

Sc. What generall note is there to be obserued, when we measure a thing at one station?

Ma. You know this, that the measuring of a thing by the Iacobs Staffe, dependeth wholly vpon proportions, whose drift and finall end is this, by thrée numbers giuen, to find out the fourth. Two of these numbers are alwaies the partes of your Staffe, the one being the partes of your Transame, the other of the yard: the which numbers must be placed according as the rules which follow shall direct you. The thirde number of proportion is some assigned measure in this manner. If you measure the length of a thing, some certaine height must be assigned. If you measure the height or breadth of a thing, some certaine length, or distance must be giuen you for your thirde proportionall number, and this number must kéepe the third place in the rule of proportion, or Golden rule, or rule of thrée, as we do commonly terme it. Then must we say thus: that, as the partes of the Transame cutte off by the Yard, are vnto the parts of the Yard cut off by the Transame: Or contrariwise: As the parts of the yard cut off by the Transame, are vnto the parts of the Transame cut off by the yard, so is the assigned measure which is giuen you, vnto the measure of the thing sought for. This is the chiefest note in measuring at one station.

Sc. What note is there to be obserued in measuring at two stations?

Ma. There are thrée things specially to be noted: The first is, the making of your stations, that they be both of them in a right line with the thing to the which you leuell your staffe, whether you go forward or backward.

Sc. How may that be done?

Ma. Chose out some marke in the thing to be measured, which may be leuell with your eye, then set vp two staues where you minde to make your two stations, so that standing at the further Staffe, you may bring the other iust with the marke which you doe chose out: so shall you be sure that the two stations are in a right line with the marke chosen. For it is a generall rule, that if three pointes be in one leuell of the eye, those three pointes are in one right line. The second is, that your eye be no higher at the one station then it is at the other. This is to be done by the rule which I gaue you euen now For if you, standing at your second station, and holding your Staffe leuell from your sight, can see the top of the Staffe which standeth at your first station, and also the marke which you chose, it is a certaine signe that your eye is in his iust height. The third note is, that the distuance betweene the two stations be diligently taken, and that in a right line, so neare as is possible: because the whole art of measuring by the Iacobs staffe, dependeth vpon a right line. This distance must be the third number of your proportion. And these be the chiefest obseruations that I remember. If there be any thing else, I will put you in minde thereof as occasion serueth.

Sc. Let vs come now vnto the particular dimensions, and beginne first with the length? What call you the lenth of the thing measured?

Ma. It is the distaunce betweene the place whereon you stand, and the place assigned, vnto which you direct your sight: and in this case we commonly say thus: How farre is it from my foote, to such, or such a thing? as a Tree or House. &c.

Sc. Giue me now the rules whereby I may direct my selfe in measuring any distance giuen.

Ma. The first rule requireth, that the yard should be perpendicular to the length measured: And that is this. The yard of your Staffe being perpendicular vnto the thing measured. If your sight passe from the top of the yard ouer the end of the Transame to the farther end of the thing propounded, as the partes of the yard cut off by the Transame, shall be vnto the parts of the Transame cut off by the yard, so shall the height of him that measureth bee vnto the length of the thing propounded.

Sc. Here are two questions to be propounded. First, what call you the end of the Transame?

Ma. That is generally called the end of the Transame, ouer the which your eie sight passeth wheresoeuer it be, whether it be in the very end of the wood it selfe, or any other place, where you list to set the Vane, so that your eye sight passe by that place.

Sc. Secondly, what call you the height of him that measureth? For some man may be, and is a great deale higher then another.

Ma. The height of the measure is the distaunce betwéene the top of the Staffe and the thing measured: as in the 1. figure from y. to o. and from a. to o. is the height of the measurer: so that whether he stand vppon the ground at the one end of the distance assigned, or whether he stand vpon some higher place, the maner of worke is all one: this being noted as a generall rule: That the height of the top of the yard from the ground, must be certainely knowen.

Sc. Expresse the meaning of your Rule by some example.

Ma. Put case the partes of the Yard cut off by the Transame bee 60. and the partes of the Transame cut off by the Yard be 180. let the height of the Staffe from the ground be 4. foote: then I say that as 60. which are the partes of the Yard, are vnto 180. which are the parts of the Transame: euen so is 4. féete, which is the supposed height of him that measureth vnto the length propounded. So that I conclude the length thereof to be 12. foote.

Sc. This is easily concluded, and dependeth, as I remember vpon the 4. p. of the 6. booke of Euclide: where it is sayed that; In equiangle triangles, the sides which containe the equall Angles are proportionall, and the sides which are subtended vnder the equall Angles, are of like proportion. Moreouer I perceiue (as you told me before) that the third number of the proportion, is the aſſigned height.

Ma. It is true: and you may moſt manifeſtly perceiue that in the Figure which you ſayd concerning Euclides propoſition: For the 2. Triangles a. e. i. and a. o. u. are like, becauſe they are ſituated after one and the ſelfe ſame maner: therfore in the Triangle a. e. i. as a. e. is vnto a. i. ſo in the Triangle a. o u. a. o. is vnto o. u. Thus you ſée the commodity of the Staffe in meaſuring the length of a thing. And here you muſt note this, That it is no matter whether the thing meaſured do lie in a plaine or flat, or whether it be in the aſcention or deſcention of an Hill, or whether there be any concauity or conuexity betweene the ends of the thing meaſured. So that by this rule, we may meaſure the breadth of any Riuer, or of any Trench, or of any Valley betwéene the tops of the hils, or the diſtaunce of two Ships at ſea &c.

Sc. The firſt Rule required, that the yard ſhould be perpendicular vnto the thing meaſured. Put caſe the yard do lie parallele vnto it, as it is in the 2. Figure, what then?

Ma. The yard being parallele vnto the thing meaſured, if your ſight paſſe from the beginning of the yard ouer the ende of the Tranſame, then, as the parts of the Tranſame cut off by the yard, are vnto the parts of the yard cut off by the Tranſam, ſo is the height aſſigned vnto the length propounded to bee meaſured.

Sc. So then by your Rule I conclude, that if the partes of the Tranſame cut off by the yard be 120. parts: and the parts of the yard cut off by the Tranſame be 210. and the height aſſigned and knowen be 400. féete, the length to be meaſured muſt be 700. féete, as I perceiue by the Golden rule, and by the propoſition of Euclide mentioned before.

Ma. You ſay very true: for conſidering the two Triangles a. u. o. and a. i. e. are like, it muſt néedes follow, that as in the leſſer Triangle o. u. is vnto u. a; ſo in the greater e. i.

muſt

must be unto i. a.

Sc. You sayd heretofore, that there were 3. waies to finde out the length of any thing, whereof 2. are already set downe: What is the third way?

Ma. The two waies already set downe, require but one station. For you are not bound to remoue from the place whereon you first set your foot when you beginne to measure. But this third way requireth a double station, as in the 3. Figure, which is many times necessary, because that a Trée, or a Wall, or an Hill, may be betwéene our sight and the end of the thing, whose distaunce from us we desire to know, so that our sight cannot come unto the end thereof. The third way differeth from the former in this, that there is no altitude to be assigned and giuen for the third number of the proportion, but the third number of the proportion, is the distaunce betwéene the two stations.

Sc. I pray you answere me to this: Considering that I must take 2. stations, at which is it best to begin? Is it best to begin my first station at the end of the distance propounded, and so go backeward? or is it best to begin my first station a good way from the end of the distaunce propounded, and then come forward to make my second station?

Ma. That is neither here nor there: But the common use is to begin the first station, at the end of the distance propounded, and so go backward: in the which thing doing, you must take héede of that especially, which I told you before, that you goe backeward in a straight line: so that the line imagined to be drawen from the one end of the distance propounded unto the other, being continued unto the second station must make but one right line: otherwise you shall erre in your measure more or lesse.

Sc. What distaunce is best to be made betwéene the two stations?

Ma. The larger the better: For it is good neuer to take a short station, if you may haue a large one: because your errour will be the lesse thereby.

Sc. Giue me now the rule which I must follow.

Ma. The Yard being parallele vnto the length propounded to be measured, and your eye at each seuerall station, passing from the head thereof ouer the end of the Transame, vnto something that standeth vpright, at the farther end of the distaunce prounded: as the difference of the greater segment of your Transame shall be vnto the lesser segment: so shall the length betweene the two stations be vnto the distance propounded.

Sc. In this kinde of measuring, must the Transame kéepe one place vpon the yard, without pulling it nigher, or putting it farther off from the sight?

Ma. The rule that I gaue you euen now requireth that it should do so. And then you must note this, that the farther you go backeward from your first station, the lower must the Transame sinke into his socket.

Sc. Giue me an example of that which you haue sayd.

Ma. Put case that at your first station the partes of the Transame cut off by the Yard were 108. at your second station, put case the partes of the Transame cut off by the Yard were 72. imagine also the distance betwéene your two stations were 40. féete: I must séeke out the difference betwéene the seuerall partes of the Transame cut off by the Yard: which by subduction I finde to be 36. (for that is the remainder, 72. being subducted out of 108.) whereupon I conclude thus, that as the difference of the greater segment of the Transame, namely, 36. is vnto the lesser segment, which is 72. so is the distance betwéene my two stations, to the length propounded: and therfore the length propunded to be measured is 80. féete. For if 36. giue 72. then 40. féete (which is the distance of the two stations) must néeds yéeld 80. Thus haue you hitherto séene the thrée seuerall waies how to measure the length of any thing propounded. It followeth now to instruct you in taking the height of any thing propounded.

Sc. What call you the height of a thing?

Ma. The height of a thing, is the perpendicular drawen from the top thereof, vnto the base whereon it standeth.

Sc. Then you regard not at all the crookednesse of a thing.

Ma. No: I care not at all for the fashion thereof. For whether it leane forward or backeward, to the right hand or to the left, I care not, so that from the top thereof I may draw down a perpendicular line: as you see in the fourth figure.

Sc. Do you take all this line?

Ma. Your measure taketh no more then is from the leuell of your eie vpward: and therfore this must be a generall rule, that, if the thing propounded to be measured, doe stand leuell with your feete, you must adde vnto the height found out, the height also of your owne stature.

Sc. How many waies is the height of a thing measured?

Ma. Three maner of waies, euen as the length or distance was.

Sc. What must bee the third proportionall number in the first two waies?

Ma. As in the measuring of lengths, the height assigned and giuen, was the thirde proportionall number: so in the measuring of heights, the length assigned is the third number of the proportion.

Sc. Giue me the rules to direct me by.

Ma. The first rule is this: Your Transame making a right angle with the height to be measured: and your sight passing from the beginning thereof to the toppe of the height: as the partes of the Transame cut off by the Yard, are to the parts of the Yard cut off by the Transame, so is the length giuen vnto the height.

Sc. Put case then the parts of the Transame were 60. and the partes of the Yard 36. and the length giuen 20. feete: the height according to the rule must be 12. feete.

Ma It is so: But you must remember to adde the height of your owne stature: so that it shall be in all 16, feete, or 15. feete according as you ſhall pleaſe to allow for your owne height

Sc. Let me requeſt this by the way: is there no rule for the meaſuring of depths?

Ma. It is all one with the meaſuring of heights. For the depth is nothing elſe but a rehearſed height. And therefore in the meaſuring of the depth of a thing, The Yard of your ſtaffe being parallele vnto the ſide of the thing, whoſe depth you meaſure, and your ſight paſſing from the toppe thereof: as the parts of the Tranſame cut off by the Yard, are to the parts of the Yard cut off by the Tranſame, ſo is the aſſigned length vnto the depth of the thing.

Sc. I perceiue your meaning. As for example: If the parts of the Tranſame be 5. and the partes of the Yard 13. and the diameter of the well (which is here in this example the aſſigned length) be 10. féete long, the depth muſt be 26. féete.

Ma. It is true, if you account from the top of your Staffe, and therefore that you may haue the iuſt depth of the thing, you muſt take away from the aforeſaid number as much as is aboue the thing meaſured: which if you ſuppoſe to be one foote in length, then is the well 25. foote déepe, as appeareth in the fifth figure.

Sc. Thus much by the way concerning the meaſuring of any depth. What other rule haue you for the meaſuring of heights?

Ma. The firſt rule required that the Tranſame ſhould be perpendicular vnto the height propounded. The ſecond requireth the Yard to be perpendicular: as in the ſixt figure: ſo that it differeth from the other onely in the manner of holding the Staffe. The rule is this: The yard of your Staffe being perpendicular vnto the thing meaſured as the parts of the yard are to the parts of the Tranſame, ſo is the aſſigned length vnto the height propounded to be meaſured.

Sc. So that I may conclude thus. If the parts of the Yard be 60. and the parts of the Tranſame 60. and the aſſigned length be 250. féete: the height propounded muſt alſo be 250. féete.

Ma. Your

Ma. Your conclusion is allowable: But I must put you in minde of this, to adde the allowance of your stature, which if you suppose to be 4. feete, then is the whole height 254. feete: Now marke this one thing for your further learning.

Sc. What is that?

Ma. By this meane you may finde out the severall partes of any height propounded: as in the seventh figure you may finde out how many feete are contained betweene the top and the nether side of the window: or generally betweene any parts else, or placing as many Vanes upon the Transame, as you would measure severall parts.

Sc. I must so place my Vanes, that from the end of the yard I may see each part propounded.

Ma. You say right, the rule then must be this: The Yard being perpendicular, as it was before vnto the height propounded, as the parts of the yard cut off by the Transame are vnto the distance of the Vanes, so is the whole height, vnto the part thereof propounded.

Sc. Let me shew that I understand you, by propounding an example: I say, that if the partes of the Transame be 50. and the distaunce of the Vanes be 20. partes: and the whole height of the thing measured be 37. feete, then the part propounded must be 14. feete, and $\frac{4}{5}$.

Ma. You understand it well. And here is yet another thing to be noted: that if you know the length of any one peece of the thing whose height you desire to know, it is sufficient to bring you to the knowledge of the whole height, as in the 8. figure.

Sc. What rule have you for that to direct me in the worke?

Ma. The Vanes of your staffe must be placed as they were before, namely. in such sort that your sight passing from the end of the yard by the Vanes, you may see both the ends of the part whose length you know. The rule then is this: that as the distance of the two Vanes is vnto the remnant of the Transam downe vnto the yard, so is the part knowen vnto the remnant of the height propounded. So that if the distance of the Vanes be 20. parts, and the remnant of the Transame from the nethermost Vane downeward be 30. parts: and the part of the height knowen be 14. feete, and $\frac{4}{5}$. I conclude the remnant of the height to be 22. feete, and $\frac{1}{5}$.

Sc. That standeth with reason, and may be proued by the example next going before this. For if the whole height were 37. feete, and the part propounded 14. feete, and $\frac{4}{5}$, that part being taken out of the whole, declareth the remnant to be 22. feete, and $\frac{1}{5}$.

Ma. The former two rules measured the height of a thing at one station. The third rule requireth a double station, as in the ninth figure. The rule is this: The Yard of your Staffe being perpendicular to the thing measured, as the difference of the two segments of the yard is to the distance betweene the two stations, so are the partes of the Transame cut off by the Yard to the height propounded.

Sc. What meane you by the difference of the two segments of the Yard?

Ma. You may easily know what that meaneth, by that which hath beene taught you before, in measuring the distance or length of a thing at two stations. Yet I will make it plaine unto you here. When you make your first station, you know the Transame must cut the Yarde in one place or other: also when you make your second station, the Transame must cut the Yard in another place, not the same which was before: the distance or parts of the Yard, contained betweene those two severall places, are called the difference of the segments.

Sc. In this case I must not moue my Transame up and downe.

Ma. No, but you must moue it only either toward or from your sight.

Sc. When must it come toward my sight? and when must it goe farther from it?

Ma. That falleth out according to your severall stations. If in making your second station, you goe farther from the thing

thing measured, your Transame must goe farther from your sight: If you come nigher, your Transame also must come nigher your sight. And here I giue you warning againe, that you goe forward and backward in a right line, directly vppon the thing propounded.

Sc. Let me here also by example shew that I vnderstand your Rule. You say that as the difference betwene the two segments of the yard is vnto the distaunce betwene the two stations, so are the partes of the Transame vnto the heigth. Whereupon I conclude thus: that if the difference of the two segments be 23. and the distaunce of the two stations be 30. feete, and the parts of the Transame cut off by the yard be 44. then the heigth must be 57. feete, and $\frac{9}{23}$. as in the Figure.

Ma. Hitherto you haue learned how to measure the length and the height of things: now it followeth to measure the breadth.

Sc. Giue me leaue, I pray you, in this place to moue thrée questions: and first what meane you by the breadth?

Ma. The breadth here to be measured, is no more, then the length of the right line, conteyned betwene the two ends of the thing propounded, without any consideration of the roundnesse or hollownesse of the thing.

Sc. Secondly, how must I stand to hold my Staffe?

Ma. You must stand so, that your staffe being applied vnto your eie, it may be perpendicular vnto the bredth propounded.

Sc. Last of all, when I purpose to measure the breadth of any thing, as of a wall, or such like, against which part of it must I stand? against the iust middest of it? or against the one end? or else where I please, betwene the two endes? or without both the ends?

Ma. It is at your owne choise.

Sc. In measuring the former dimensions, my sight did passe but by one end of the Transame, must it do so likewise in this?

Ma. That is according as you stand to take the measure.

If you stand within the two ends of the thing propounded to be measured, your sight must passe by the two ends of the Trãsame. If you stand against the one end of the thing, or cleane without it, your sight passeth but by one end of the Transame.

Sc. What meane you here by the ends of the Transame? are they not here also to be taken as they were before?

Ma. Yes: The one end of the Transame must alwaies be that, where the partitions begin: the other end is that, vppon which the Vane must stand, and by which the sight passeth. For many times it will so fall out that your sight cannot passe by both the ends of the wood, so that you shall be enforced to place the Vane in some one part or other of the Transame, that it may be a limit for your sight And this is a generall rule, that if you stand within the ends of the breadth propounded, the one end of your Transame must be on the one side of the yard, and the other on the other side.

Sc. But will the yard be iust betwéene them both?

Ma. No: Except your station be made accordingly. If you stand in the middest of the thing to be measured, then will the yard be iust in the middest, betwéene the endes of the Transame: if you stand more to the one end then to the other, the yard also will fall out in like manner, more toward the one end of the Transame, then toward the other.

Sc. How many waies are there to be vsed to measure the breadth by?

Ma. We may measure it two waies: either by the helpe of one station, or of two.

Sc. If I measure it by the helpe of one station, what must be knowen for the third proportionall number?

Ma. The distance betwéene you and the thing propounded.

Sc. Put case that it be hollow inward betwéene the two ends thereof: or swell outwardly: to which part must I then measure?

Ma. If it be conuexe, or round outward, the best way to measure the breadth is by two stations: but if it be hollow inward

ward, you must measure no further then from your féet to the right line, which runneth betwéene the two ends of the thing propounded.

Sc. What rule is there for the measuring of a breadth at one station?

Ma. This. The Yard being perpendicular vnto the breadth assigned, as the parts of the Yard cut off by the Transame are vnto the partes of the Transame contained betweene the two ends, so is the distance betweene your feete and the thing measured, to the breadth thereof. As in the tenth figure.

Sc. I perceiue it well: For there are here two proportionall triangles cut in péeces with a line in the midst: Therefore as in the triangle A B C. the line B C. is to the line A C. so is the line B F. in the greater triangle to the line F G. which is the breadth of the wall. This dependeth vpon that which went before. Now what is your rule for the measuring at two statiõs.

Ma. It is this: The Yard being perpendicular vnto the thing measured, & your sight passing by the ends of the Transame (as it is required:) as the difference of the two segments of the yard is vnto the distaunce betweene the two stations, so are the parts of the Transame betweene the two ends thereof, vnto the breadth propounded: as it is in the 11. figure.

Sc. This also is very easie by reason of that which was taught before. So that I may conclude, that if the difference of the two segments in the yard be 30. parts: and the distance of my two stations be 40. féete: and the parts betwéene the two ends of the Transame be 50. then must the bredth of the thing be 66. féete and $\frac{2}{3}$. Haue you any other thing to be added?

Ma. No: But that I would haue you to practise this diligently, and I will hereafter further you in other things.

Ma. I thanke you for it. Yet before I depart, I pray you let me aske you one question: You said before, that in measuring the breadth of any thing it is at my choise to stand where I will, either within the two ends of the thing to be measured, or against the one end thereof, or without both the endes, in what place I shall thinke most expedient. This last kinde of measuring me thinketh is strange (as for the other I perceiue well enough how their proportion will fall out) wherefore I pray you helpe me herein.

Ma. Set the question as you shall thinke it best for your owne vnderstanding.

Sc. Put case I were to take the breadth of a riuer, as in the 12. figure, vnto whose brinke I can not come any nigher then 40. féete, by reason of some marish grounds adioyning therevnto: what meanes shall I vse to finde the breadth thereof?

Ma. You must goe thus to worke: (and as you do in this, so must you doe in all other things, without whose ends you bée inforced to stand when you take their breadth) chose out vpon the riuers side two markes, as two trées, or such like things, standing one right against another.

Sc. But how must they stand in respect of the riuer?

Ma. Indéed that is a speciall thing to be noted: They must stand so, that the right line imagined to come from the two markes, may make a right angle with the riuers side: otherwise you shall erre in taking of your measure: Because that only is accounted the breadth of a thing, which maketh a right angle with the length.

Sc. Hauing chosen out these two markes, what is then to be done?

Ma. You must set vp another, which must stand also iust in one right line with the two first markes, whose distaunce from the mark on this side the riuer, must either be giuen you, as being already certainely knowne: or else it must be found out by the first rule concerning the measuring of breadths.

Sc. Put case I either know it, or finde it to be 40. féete.

Ma. You must chose out a conuenient place, wherein you may make your station, & settle your selfe about your worke.

Sc. How must that place lie from the marke that was last set vp?

Ma. It must lie from it so, that the line drawne from that

marke

marke vnto the place of your standing, may make a right angle with the line drawne betweene the thrée aforenamed markes.

Sc. Must I not know the iust length betwéene the place wherein I make my station, and the marke which I set vp.

Ma. Yes: For it is the third proportionall number.

Sc. Imagine that distance to be 60. féete.

Ma. Then your rule to finde out the breadth is this: The Yard being perpendicular with the marke set vp, and your sight passing from the head thereof by the end of the Transame vnto the marke standing on the farther side of the riuer, as the parts of the Yard cut off by the Transame, are vnto the partes of the Transame cut off by the Yarde: so is the distaunce betweene your station and the marke set vp vnto the distaunce betweene the sayd marke, and the marke standing vppon the further side of the riuer.

Sc. But what is this to the question that I propounded? you make an aunswere concerning the distaunce of the two markes, whereas I moued a question of the breadth of the Riuer.

Ma. Be not angry, I will satisfie you straight waies: first finde out the distance of these two markes.

Sc. That can I easily doe: For the parts of the yard being 30. and the parts of the Transame 36. and the distance of my station from the first marke 60. féete. The first marke must be distant from the third 72. féete.

Ma. Then what is the breadth of the riuer?

Sc. That must come from you, for I cannot make the conclusion.

Ma. Then marke this. You sayd the distaunce of the first marke from the third being on the farther side of the Riuer is 72. féete: and the distaunce of the first marke from the second being on the hither side of the Riuer, was graunted to be 40. féete: then take 40. out of 72. and the remainder being 32. féete, declareth the breadth of the Riuer.

Sc. I perceiue now what a commodity it is to haue an instructor. For the thing which seemed vnto me most strange, and altogether impertinent to my question, is by your meanes made most euident vnto me, and greatly to the matter propounded. But must I take this course, whensoeuer I am driuen to stand without the compasse of the thing measured.

Ma. Yea, that you must: And sometimes also you shall be driuen to the second rule, wherein two stations are required, which soeuer you vse it commeth all vnto one purpose, and will easily afford you the breadth of a thing.

Sc. I can but thanke you for your paines, and good will, in accepting me for your scholler.

Ma. And I likewise you, for making choise of me to be your maister. My will is good to helpe whom I may: howsoeuer others may make more boast of their cunning.

FINIS.